Nature's Children

SEA MAMMALS

by Frank Puccio

GROLIER
EDUCATIONAL

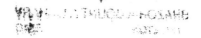

FACTS IN BRIEF

Classification of dolphins, porpoises, whales

Class: *Mammalia* (mammals)

Order: *Cetacea*

Suborder: *Odontoceti* (toothed whales, dolphins, porpoises)

Families: *Delphinidae* (dolphins); *Phocoenidae* (porpoises); *Physeteridae* (sperm whales); *Monodontidae* (white whales); *Ziphiidae* (beaked whales)

World Distribution: In all seas and certain lakes and rivers.

Distinctive physical characteristics: Live under water; have fins and horizontal tails; breathe through a blow hole on top of their heads.

Classification of true seals, fur seals, sea lions, and walruses

Class: *Mammalia* (mammals)

Order: *Pinnipedia* (pin-haired animals)

Families: *Otariidae* (fur seals and sea lions); *Odobenidae* (walruses)

World Distribution: From the Arctic Ocean to Australia, Africa, and South America.

Distinctive physical characteristics: Have flippers for limbs; rough fur coat; true seals and sea lions have outer ears, fur seals do not; fur seals have a soft undercoat; walruses have almost no hair, no outside ears, a mustache, and two tusks.

Library of Congress Cataloging-in-Publication Data

Puccio, Frank, 1956-
    Sea mammals / Frank Puccio.
      p.   cm. — (Nature's children)
    Includes index.
    Summary: Describes the physical characteristics, behavior, and natural habitat of whales and dolphins, while focusing on the cetaceans that might be encountered in a zoo or aquarium.
    ISBN 0-7172-9121-9 (hardbound)
     1. Cetacean—Juvenile literature.   [1. Whales. 2. Dolphins.]
I. Title.   II. Series.
QL737.C4P83 1997
599.5—dc21   97-5946
                                                       CIP
                                                       AC

This library reinforced edition was published in 1997 exclusively by:

Grolier Educational

Sherman Turnpike, Danbury, Connecticut 06816

Set ISBN 0-7172-7661-9
Sea Mammals ISBN 0-7172-9121-9

Contents

Sea Mammals?   Page 6

Special Equipment   Page 9

Dolphins and Porpoises   Page 10

Capturing Dolphins   Page 12

How Smart Are They?   Page 15

Sounding Off   Page 16

Training: The Basics   Page 18

Training: More Basics   Page 19

Hitting a Target   Page 21

Learning Complex Tricks   Page 22

You're in the Navy Now   Page 24

Danger Ahead   Page 27

Whales   Page 28

Whale Behavior   Page 30

Whale Intelligence   Page 32

Deadly Entertainers   Page 34

Movie Stars   Page 37

Really Freeing Willy   Page 38

Rescuing Whales   Page 40

Protecting the Whale   Page 42

Seals and Sea Lions   Page 43

The Walrus   Page 44

The Future of Sea Mammals   Page 46

Words to Know   Page 47

Index   Page 48

In captivity sea mammals have become great entertainers
and have been taught to do complicated tricks.

They would seem impossible to domesticate, or tame.
They are simply too large, too swift, and too different
from the other animals that inhabit the earth. Still,
they star in movies and perform complicated tricks.
They find lost equipment far beneath the sea, and
they help scientists study the deepest mysteries of
communication and intelligence.

What are they? They are sea mammals—dolphins
and porpoises, whales, seals, sea lions, and walruses.

Over the years we humans have watched and
admired these amazing creatures. We also have
depended on them for food, clothing, heat, and light.
And carelessly we have hunted some of them almost
to extinction.

Today concern for the future of sea mammals is
growing. Much of this awareness is due to what
people see in films and in marine parks. The public,
it seems, just cannot get enough of these creatures.
As a result, more and more sea mammals are being
brought into environments created and maintained
by people. But even more important, the positive
feelings surrounding these creatures mean that there
is a better chance for them to survive.

What Are Sea Mammals?

Sea mammals come in all sizes. Some, like seals, are fairly small. Others, such as whales, are among the largest creatures on earth. Some—dolphins, porpoises, whales—usually live far out at sea, while seals, sea lions, and walruses live close to shore and on land. Regardless of their size or where they live, these creatures seem at home in the water. It is there, as they dive and leap and dart about, that their grace and speed are shown for all of us to admire.

What are sea mammals like? First of all, like all mammals, they are warm-blooded, which means that they maintain their own body temperature regardless of the temperature of their environment. Also, mammal's babies are born alive, not in eggs, and are fed on milk that is made within their mothers' bodies. Finally, unlike fish, mammals have lungs, organs that allow to them breathe air.

These creatures, however, are so well suited to life at sea that some of them—dolphins, for example—are even born underwater. Beyond this sea mammals are shaped to cut cleanly through the water. Instead of legs or arms they have tails and flippers to move them through the water as swiftly as any fish. They also have fur or extra layers of fat to keep them warm.

Some sea mammals live almost entirely in the water, while others— including seals and sea lions—spend much of their time on land.

Special Equipment

Unlike fish—which breathe through gills—mammals get the oxygen their bodies need from air. This means that they cannot breathe underwater but must come to the surface for air.

Seals, sea lions, and walruses breathe, as we do, through their noses and mouths. This is convenient for them since they spend so much time on rocks or beaches.

Dolphins and whales, however, spend almost all of their time in the water. Their breathing is done through blow holes located at the very tops of their heads.

Dolphins and whales have other special equipment as well. Because it is so dark beneath the sea, they cannot rely on eyesight to guide them as they swim and look for food. Instead, they have a built-in system, called echolocation, that is much like the sonar instruments used on ships to locate objects underwater.

The dolphin or whale sends out sounds that echo, or bounce back, when they strike a fish, a rock, or any other object. A special part of the animal's brain then uses that echo to figure out where the object is located.

Dolphins and porpoises breathe through blow holes on the tops of their heads.

Dolphins and Porpoises

Dolphins are among the world's most popular sea mammals. There are several types, but the best known is the bottle-nosed dolphin, which is the animal we see entertaining visitors at marine parks and sea shows.

In the wild bottle-nosed dolphins are found from the northwestern Atlantic Ocean to the Mediterranean Sea to the Pacific. Atlantic bottle-nosed dolphins grow to between 8.2 and 9 feet (2.5 to 2.7 meters). Pacific and Mediterranean ones reach 12 feet (3.7 meters) and more.

Although their colors often differ a bit, dolphins usually have light-colored bellies and dark backs. These colors help the animals blend in with their surroundings. As they swim, their backs are almost invisible against the dark ocean surface. From below their light bellies are hard to see against the bright surface waters.

Everything about the dolphin, from skin to fins, helps it move through the water. Dolphins even have a special substance that keeps their eyes moist and clean underwater.

What's the difference between a dolphin and a porpoise? The truth is that they are so similar that it is easy to confuse which is which. In general, however, porpoises are smaller and plumper than dolphins and dolphins have a beak-like snout.

Both dolphins and porpoises leap gracefully as they move through the water.

Capturing Dolphins

Great care must be taken when dolphins are brought into captivity. First, nets are used to bring the animal close to the boat that will bring it ashore. All the while swimmers stay near the animal, keeping it from tangling itself in the nets or hitting the sides of the boat.

Then a special stretcher is used to lift the animal out of the water. There are holes for the animal's flippers and special protection for its eyes. On deck the animal is placed on foam pads. This protects the animal's internal organs and keeps its blood moving smoothly.

Out of the water the dolphin must be kept moist. Otherwise its skin will dry, and the dolphin will overheat.

In the past handlers simply poured buckets of water over the animal. Today they use moisturizing creams to form an insulating layer over the skin. Blankets are then put over the dolphins and kept wet with buckets and sprinklers.

Dolphins can be taught many clever tricks.

Dolphins seem to enjoy working with people—and doing tricks for them.

How Smart Are They?

Have you ever seen the movie *Flipper* or watched dolphins perform at an aquarium or marine park? Do you wonder how these animals manage to do such clever tricks?

The easy answer is that they are smart. After all, an animal must be smart in order to master complicated tricks or to respond to people in such a lively way.

The truth, however, is more complicated. As with other creatures, it is hard for humans to judge just how intelligent dolphins and other sea mammals might really be. Still, we do know that sea mammals—dolphins and whales in particular—have large brains. We also know that these creatures can be trained to follow difficult commands and to do tasks that take two, three, or even several steps.

Scientists have also found that sea mammals learn at different rates. Some species learn faster than others, and some individuals seem to learn more quickly than other members of the same species. Unfortunately, until more research is done, that is about all anyone can say for sure regarding the intelligence of sea mammals.

Sounding Off

Scientists may still be uncertain just how intelligent dolphins are. But few scientists doubt that these creatures are able to communicate.

Over the years people have recorded many of the sounds dolphins make. At first these recordings seem to contain just squawks, squeaks, whistles, clicks, and groans. But as scientists have examined the sounds more closely, they have found that these noises are messages from one dolphin to another. With these sounds dolphins warn each other of danger, cry for help, or even call out invitations to a future mate. More amazing, even other kinds of sea creatures seem to recognize and understand the dolphins' messages.

Dolphins also seem to have individual voices, just as people do. Young dolphins start to make sounds just a few days after they are born. Then, as they grow older, each of them develops its own "signature" sounds.

As yet no one has been able to decode fully the meaning of dolphins' sounds. But in time scientists hope to master the art of dolphin communication— and even to be able to communicate back!

Scientists have found that dolphins use their individual voices to communicate with each other— and with other kinds of creatures.

Training: The Basics

Dolphins, like other sea mammals, can be trained to a surprising level. But the training of sea mammals is a fairly new field. Because of this scientists and handlers are only just beginning to discover what these animals can learn and do.

Training these animals takes patience. It also takes an understanding of two important ideas: reinforcing and signaling.

Reinforcing is a way of rewarding an animal for a certain kind of behavior. Basically it means that if you want a dolphin to do something, you give it a reward for doing it. For sea mammals that reward often is food. But it might also be a rubdown, a squirt of water from a hose, or even a chance to play with a particular toy. What works best depends on each animal. Trainers, therefore must experiment to find out exactly which reinforcement works best with each particular animal.

Once trainers decide on a reinforcement, they use it again and again. Their goal is for the animal to link the behavior and the reward. For dolphins and other sea mammals this is a key first step in their training.

Training: More Basics

It is one thing for a sea mammal to link a particular behavior—jumping, splashing a handler, leaping through a hoop, or whatever—with a reinforcement or reward. But how does the handler let the animal know when to do that trick?

The answer involves the other key idea of training—signals. Signals let the animal know exactly when it is time to perform a certain action. The signal may be a whistle, a light touch of the hand, or even a certain word, such as "okay." Whistles, for example, are the most common signals used for dolphins. Touches and words seem to work well with sea lions.

Each time the animal does the action the trainer wants, it is given the same signal and the same reinforcing reward. This is done over and over again until the animal links all three things together—the signal, the action, and the reinforcement. After a while—if everything goes right—the animal actually will start to perform the right action whenever the right signal is given.

Hitting a Target

One of the first behaviors, or tricks, sea mammals learn is to hit or touch a target. Here's how trainers use reinforcements and signals to do it.

Handlers begin by touching the animal with the target. But each time this is done, the signal—a whistle, for example—is given. And with the whistle and the touch comes a reinforcement—a fish. Again and again this is repeated: whistle, touch, fish; whistle, touch, fish.

After a while the animal knows that whenever it touches the target, it will hear the whistle. It also knows that it will get a fish. Once the trainer is sure of this, the target can be moved a short distance away.

This time the trainer doesn't bring the target to the animal. The animal must go and touch the target. When it does, it hears the whistle and gets the fish. This is repeated over and over again until the animal goes right for the target whenever it hears the whistle.

Training continues with the target being moved, in stages, further and further away. In time the animal learns the trick. More important, it has learned something that will help trainers teach the animal more.

Trainers use a system of signals and
rewards to teach sea mammals tricks.

Learning Complex Tricks

Having a dolphin touch a target, even one several feet away, is an accomplishment. But it is not the kind of stunt that draws gasps and cheers from a crowd at a marine park.

How do trainers get dolphins to perform more complicated tricks? The answer is to break the trick down into simple, easy-to-master steps.

Suppose someone wants to train a dolphin to touch a target held high above the surface of the water. Training begins by teaching the dolphin to touch a target held right on the surface of the water. When the dolphin masters this, the target is raised a bit. Once again the dolphin and trainer follow the pattern—signal, behavior, reinforcement—until the dolphin links all three elements together.

As the dolphin masters each step, the trainer moves on to the next one. The target is raised higher and higher until it finally is at the proper height for the trick. Then, when the dolphin is able to touch it every time in practice, it is time for the real show to begin!

*By breaking complex tricks into simple steps trainers
can teach sea mammals to do amazing stunts.*

You're in the Navy Now

Circuses and marine parks are not the only organizations interested in training sea mammals. Believe it or not, even the U.S. Navy thinks that these creatures are smart enough for duty.

In 1969 the naval Undersea Center in Hawaii started an unusual project called Quick Find. Its purpose was to train sea lions to take on special missions for the navy.

Because sea lions can dive so deep and stay underwater for so long, they are perfect for search-and-rescue operations. With proper training the sea lions were soon at work finding equipment that had been lost at sea.

Dolphins have had an even more interesting navy career. Like scientists around the world the navy has long been fascinated with how dolphins communicate and how they use echolocation to identify objects underwater. Their studies have helped the navy plan its own sonar operations. There have even been suggestions that dolphins guard the harbors of our cities and coastlines.

Underwater, dolphins use echolocation to find food, avoid enemies, and navigate from place to place.

Dolphins are as beautiful as they are fascinating.

Danger Ahead

The future of dolphins and porpoises is not without its dangers. As a matter of fact, thousands of these creatures die each year because of people's careless behavior.

In the last 30 years almost 7 million dolphins have drowned after having been caught in tuna nets. Commercial fishing fleets are under great pressure to capture fish quickly and economically. As a result, many use large nets to capture tuna and other fish. When a fishing boat discovers that a school, or group, of tuna is nearby, it sets out its nets. As the school of tuna goes by, fish are caught in the nets.

Unfortunately, dolphins are also caught. Trapped in the nets, the dolphins are hauled through the water, unable to get to the surface for air. Eventually they drown.

In recent years protests against this type of fishing have led some tuna companies to abandon these netting procedures. Some are even advertising that their fishing boats are "dolphin safe" and do not kill dolphins and other sea mammals.

Whales

Whales are among the largest and most amazing creatures on earth. Able to dive to depths of more than 3,500 feet (1.1 kilometers), some, like the giant sperm whale, can stay underwater for as long as 75 minutes.

Whales also come in many sizes, from the relatively small belugas and narwhals to the giant sperm and blue whales. The blue whales are the largest creatures to ever live on earth, weighing up to 190 tons (172 metric tons) and reaching lengths of 100 feet (30.5 meters) and more.

Many whale species roam our seas. Among the most famous is the sperm whale, which has been hunted almost to extinction by whaling ships. The humpback whale is also well known. People sign up by the thousands for day-trips on the ocean in hopes of seeing a humpback make its leaps and jumps in the air.

Among the more unusual-looking whales is the beaked whale. With tiny fins and flippers, and pink or white blotches all over its body, it makes a strange sight. Making it appear even more odd are two teeth that stick out of its lower jaw like a pair of horns!

*The leaps and jumps of a whale
are an astounding sight.*

Whale Behavior

In the wild whales often travel in groups that are called schools, herds, pods, or gams. Some groups get together only when the whales are feeding or when they are migrating (traveling). But some whales—sperm whales, for example—form groups that stay together for as long as ten years. A few whales—beaked whales in particular—are less social. Beaked whales are rarely seen in groups of more than three.

Whales are not only social. They also are quite playful. Pilot whales will swim on their backs or even rest hanging straight up and down in the water. Sometimes they'll do this with their tails sticking straight up in the air!

Whales have also been seen swimming together, with their flippers touching as if they were holding hands. Some whales will even ride in the waves formed by ships as they plow through the water. The whales, it seems, like the humming sound of a ship's engine!

The orca, or killer whale, is just one of many fascinating kinds of whales.

Whale Intelligence

Whale's brains are in proportion to their bodies, which means that they are large . . . very large. The average size of an adult sperm whale's brain is about 20 pounds (9 kilograms). In contrast, a human brain averages only about three pounds (1.4 kilograms). The physical size of an animal's brain, of course, is not always an accurate measure of a creature's intelligence. But whales do have an unusual ability to learn. They also can be taught to perform a variety of tasks and tricks.

Like dolphins, whales seem to be able to communicate. After many scientific studies and experiments, scientists are beginning to understand that the sounds whales make—whistles, calls, and so on—most likely carry messages. A Canadian researcher even maintains that each pod of killer whales has its own particular way of "speaking," a kind of accent, just as people from different parts of a country have their own accents and ways of speaking.

Belugas are common—and appealing sights—in aquariums and marine parks.

Deadly Entertainers

One of the most popular sights at any marine park is the orca, commonly known as the killer whale. These black-and-white creatures can reach 11 tons (9.9 metric tons) in weight and lengths up to 32 feet (9.8 meters).

In the wild these creatures travel in pods, just as most whales do. But unlike other whales, the orca is a predator, living on everything from fish and squid to birds, seals, and even dolphins and porpoises. Fierce enough to kill a sea lion and strong enough to leap out of the water with the sea lion still in its mouth, an orca is a violent creature. Pods of orcas even work together to attack a good-sized whale.

Strangely, however, orcas can also be quite gentle. Among themselves, when they are not hunting, they are loyal and peaceful. In captivity they also can be trained and trusted to behave well among humans. Perhaps it is this—the idea of a fierce predator performing tricks in an aquarium—that makes orcas such popular attractions.

Fierce in the wild, orcas are gentle and trainable in captivity.

Movie Stars

In 1993 a major motion picture company did the unthinkable. It made a killer whale into a movie star.

Over the years audiences have gotten used to animals in movies. The story of the dog Lassie, for example, began as a book and then progressed through a series of movies, a TV series, a second TV series, and then another movie. Horses have been featured in such classic films as *National Velvet* and *The Black Stallion.*

Somehow, though, making a 10-ton sea creature into a hero seemed to be a major challenge. But in *Free Willy* the movie-makers did just that. This story of an orca, its human friends, and the attempt to send it back into the wild became one of the great hits of the year.

Willy was portrayed by an orca named Keiko who had long been the main attraction at an amusement park in Mexico City, Mexico. As the result of Keiko and the movie, public sympathy for the problems facing whales and other sea mammals grew even more.

Orcas are capable of crowd-pleasing stunts and tricks.

Really Freeing Willy

The story of Keiko, however, did not end with the movie *Free Willy*. After the movie was finished, Keiko went back to life at the amusement park in Mexico City.

It was not long before shocking stories began to be heard. Keiko, people declared, was living in such terrible conditions that he had become seriously ill. The park owners disagreed. Keiko, they claimed, was simply suffering from a skin condition.

Fans of the movie did not accept these answers. They founded a Free Willy foundation and raised enough money to bring Keiko to the United States.

The organization's plan was ambitious. Keiko was first to be moved from Mexico City to Newport, Oregon. There, in a new saltwater tank, Keiko would get back his health. After that he would be taught how to take care of himself in the wild—hunting fish, finding a mate, and so on. And then he would be set free.

In January 1996 Keiko set out on his journey. Thousands of people bid him good-bye as his giant water tank went to the airport. But thousands also welcomed him when he finally arrived in the United States. There people cheered at the sight of the orca that someday might be really free.

Perhaps scientists will someday help people and whales communicate with each other.

Rescuing Whales

Keiko's rescue was an unusual one. But the truth is that other kinds of whale rescues go on all the time.

For mysterious reasons, whales sometimes get stranded on a beach. Perhaps their sonar-type guidance system has gone wrong, or perhaps they have followed some strange sound they mistook for other whales. But for whatever reason, these animals sometimes find themselves caught on land, doomed to die if they cannot get back into the water soon.

Years ago these creatures were left to die. In fact, in some parts of the world people rushed out to share in the meat, blubber, and other riches.

Now, however, it often is rescuers who rush to the beach, trying desperately to save the whales. The animal's vital signs—heartbeat, body temperature, and so on—are checked. Ice and water, as well as wet sheets, blankets, and towels, are used to keep the animal wet and cool. Drugs are given to fight off disease and infection. Food and liquids are given to restore its health. Then, if possible, the whale is returned to the sea. If this can't be done, the whale is taken to an appropriate facility where it can spend its days in peaceful captivity.

In spite of their size, whales are quite graceful in the water.

Protecting the Whale

Whales have been hunted since ancient times. In some cases the damage to the whale population has been light. The Inuit people—Eskimos, as they sometimes are called—have long depended on whales for food, clothing, fuel, and tools. But the number of whales they have killed has barely made a dent in the population.

By the 1800s, however, whaling had become a huge industry. By the middle of the 1900s it was clear that many whales were on the brink of disappearing from the planet. In 1946 a number of nations got together to form the International Whaling Commission. It tried to establish rules governing how many whales could be killed.

Over the years the commission has done what it could to protect the whales. But some countries refuse to obey the rules. Others simply cheat. As a result progress has been slow. Even now, years after the founding of the commission, there still is a battle to keep the whales from disappearing forever.

Seals and Sea Lions

Seals are among the most appealing of all mammals. Visitors to zoos and marine parks always seem to gather around the seals to listen to them bark, watch them play, and admire their swift movements in the water. There are two types of seals—true seals (the type of seal that has no external ears) and fur seals (seals that have ears). And both types are likely to be found entertaining visitors either with their natural antics or with tricks learned from their trainers.

In the wild sea lions, a larger cousin of the fur seal, live mostly along the shores of the Pacific. They too are readily trained, and they almost always are part of the show at any marine park.

These creatures are trained in much the same way as dolphins, orcas, or belugas. Reinforcements and signals are used to set up links with certain behaviors. Then, with hard work and practice, they can become popular performers able to do any number of clever tricks.

The Walrus

Walruses are one of the most comical-looking of all creatures. From their huge, almost hairless bodies and their small heads to their stiff bristle mustaches and enormous tusks, they never cease to amuse visitors and audiences alike.

Like other sea mammals walruses tend to be herd animals. The herd, in fact, is one of the walrus's main forms of defense. Despite their fierce-looking tusks, walruses are quite gentle in nature. But when one of them is attacked, the whole herd will rush to defend its fellow herd member.

Although they are not as common a sight at marine shows as orcas, dolphins, or sea lions, walruses can be tamed and trained. Perhaps they, like the orcas before them, simply need a blockbuster hit movie or two to make them popular favorites.

*Walruses, too, are sometimes found entertaining
enthusiastic audiences.*

The Future of Sea Mammals

Over the years scientists have studied sea mammals in captivity and in the wild. But despite all the research, these creatures still remain quite mysterious.

One of the things that has been learned, however, is that these animals are extremely complicated and sensitive. They are not simply intelligent. They seem, say many scientists, to have feelings not unlike those of people. Sea mammals seem to suffer when they are forc to live in the small space of a tank or when they no long actively hunt for their own food. More than this, they seem to go through a deep sense of loss when they are c off from their families and herd.

Such facts seem to suggest that it is wrong to keep sea mammals in captivity or that we need to improve the conditions under which captive sea mammals live. These are just some of the problems facing us about se mammals. With luck, we will find answers before it is t late—for them and for the people who love and appreciate them.

Words to Know

Echolocation A system in which an animal sends out sound and receives echoes in order to detect shapes and objects; used by such sea mammals as whales and dolphins and by such land mammals as bats.

Flippers Finlike limbs that help sea mammals swim and steer through the water.

Gam A group of whales.

Gills The organs that take oxygen from water and allow fish to breathe.

Migrating Moving from place to place over a great distance on a regular basis; many animals migrate, as birds do, at the change of seasons.

Pod A group of whales.

Reinforcing Using rewards as a way to make an animal want to perform a certain action.

Signals The use of whistles, pats, or other signs to let a trained animal know when to perform a certain action.

Sonar Short for sound navigation and ranging; a system that uses sound waves to locate objects; widely used by navies for locating underwater objects and warships.

Target An object used in training sea mammals; animals are often taught to touch a target as part of their training.

Warm blooded Having a system that maintains its own body temperature, independently of the temperature outside.

INDEX

Atlantic Ocean, 10

beaked whales. 28, 30
belugas, 28, 43
Black Stallion, The, 37
blow holes, 9
blue whales, 28

circuses, 24

dolphins, 5, 6, 9, 10, 12, 15, 16, 18, 22, 24, 27, 32, 34, 43, 44

echolocation, 9, 24
Eskimos. *See* Inuit.

Flipper, 15
Free Willy, 37, 38

gams, 30
gills, 9

Hawaii, 24
herds, 30, 44
humpback whales, 28

Inuit, 42
International Whaling Commission, 42

Keiko, 37, 38, 40
killer whale. *See* orca.

Lassie, 37

marine parks, 5, 24, 43
Mediterranean Sea, 10
Mexico, 37
Mexico City, 37, 38

narwhals, 28
National Velvet, 37
Newport, Oregon, 38

orca, 34, 37, 38, 43, 44

Pacific Ocean, 10, 43
pilot whales, 30
pods, 30, 32, 34
porpoises, 5, 6, 10, 27, 34

Quick Find, 24

reinforcement, 18, 19, 21, 22, 43
reinforcing, 18, 19

sea lions, 5, 6, 9, 24, 34, 43, 44
seals, 5, 6, 9, 43
schools, 30
signal, 18, 19, 21, 22, 43
sonar, 9, 24, 40
sperm whales, 28, 30, 32

target, 21, 22
training, 18, 19, 21, 22, 43
tuna, 27

U.S. Navy, 24
Undersea Center, 24

walruses, 5, 6, 9, 44
whales, 5, 6, 9, 15, 28, 30, 32, 40, 42

Cover Photo: SuperStock, Inc.
Photo Credits: Ann Hagen Griffiths (Omni-Photo Communications), page 23; Lynn M. Stone, pages 7, 8, 17, 45; Bruna Stude (Omni-Photo Communications), pages 11, 26; SuperStock, Inc., pages 13, 20, 25, 29, 31, 33, 35, 36, 39, 41; Wildlife Conservation Society, pages 4, 14.